# Software Configuration Management

## A How To Guide for Project Staff

David Tuffley

To my beloved Nation of Four
*Concordia Domi – Foris Pax*

*Configuration Management is at the heart of more failed projects than many a project manager would like to admit. David Tuffley*

## Acknowledgements

I am indebted to the Institute of Electrical and Electronics Engineers on whose work I base this book, specifically IEEE Std 1042.

I also acknowledge the *Turrbal* and *Jagera* indigenous peoples, on whose ancestral land I write this book.

# Contents

# Contents

# A. Introduction

Configuration management (CM) is the regulation of the way in which a software product evolves during the development and maintenance phases of the product lifecycle. It is the process by which the individual components of a software system are identified so that any changes to the configuration of these components can be systematically traced and therefore controlled throughout the complete span of the development cycle. CM thus forms the basis for product and project measurement.

This standard is based in large part on ANSI/IEEE 1042.

## A.1. Scope

This standard describes the application of configuration management (CM) principles to the management of software development projects. CM consists of two major aspects; planning and implementation.

For those planning software configuration management activities, this standard provides guidelines into the aspects which must be considered.

Those implementing software configuration management disciplines can use the sample Configuration Management Plan attached.

## A.2. Objectives

To achieve the above, the objective of a project manager will be the identification and establishment of baselines; the review, approval, and control of changes to the project components; the tracking and reporting of such changes; the reviews of the evolving product; the control of documentation; and the control of the interfaces to the clients and supplier.

However it should be noted that this Standard specifies the minimum requirements, and therefore the project manager has the option where required, to expand and supplement as necessary for the development of specific project activities.

In specific terms, the objective of CM procedures are to provide methods for.

- Version identification

- Obtaining approval to implement a modification

- Ensuring that modifications are correctly integrated through formal change control procedures

- Controlling the identification of development status

- Ensuring that nonconforming software is identified and segregated

# A.3. References

| | |
|---|---|
| IEEE Std. 1042 | Guide to Software Configuration Management |
| IEEE Std 828 | Software Configuration Management Plans (1990) |
| ST0006 | Document Production |
| ST0010 | Project Change Control Process |
| ST0026 | Project Status Reporting |

# A.4. Sources

| | |
|---|---|
| ISO 9001 | Quality systems - model for quality assurance in design/development, production, installation and servicing. |
| AS 3901 | Quality systems for quality assurance in design/development, production, installation and servicing |
| IEEE Std 1042 | Guide To Software Configuration Management (1987) |
| IEEE 1298 | Standard for software quality management systems |

5

AS 3563            Software quality management system

# A.5. Definitions & acronyms

| | |
|---|---|
| Configuration Management | The process of identifying and defining the hardware / software configuration items in a system, controlling change of these items throughout the development process, recording the status of the items and verifying the status of the items. |
| CM | Configuration Management - see above. |
| Configuration Items | The designated hardware/software components which comprise a user function. CIs vary in size and complexity. During the development stage, CIs are only those components specified in the Customer Agreement. |
| CI | Configuration Items - see above |
| PCR | Project Change Request |
| Baseline | Gives a baseline snapshot of the particular set of Configuration Items at a specific point in time. Refer Appendix I for a more detailed description. |

| | |
|---|---|
| Plan | In this document the Plan refers to the CM Plan, as opposed to other project plans. |
| Promotion | The transition to a higher level of authorisation needed to approve changes to a controlled entity. |
| Release | Indicates certain promotions of items which are distributed outside the development organisation. |
| Revision | Correction of errors (i.e. bug fixing) |
| Software Configuration Management | The configuration management of all software items, including documentation. |

## A.6. Responsibilities

These following staff have responsibilities associated with this standard.

All project managers will be accountable for performing the required activities on all development projects, as set out in this Standard.

# B. Configuration Management Procedure

CM Planning is a mandatory component of Project Planning. The CM Plan may exist either as a stand-alone document, or be placed under section in the Project Plan. With either method, the CM Plan shall contain all the planning information either by inclusion or by reference to other documentation.

## B.1. Software configuration management issues

### B.1.1. Context of software configuration management

Software configuration management is presented here as a set of management disciplines embedded in the software engineering process. It outlines the minimum standards necessary for effective software configuration management and allows room for individual methodologies which have been developed at sites to be superimposed.

# Software Configuration Management

Software configuration management and release processing are done within the context of the following basic software configuration management functions.

- Configuration identification.

- •Baseline management.

- Change control.

- Library control.

- Status accounting.

- Reviews and audits.

- Release processing.

The way in which these functions are performed varies according to the kinds of programs being developed and may vary in the degree of formal documentation required.

Software configuration management also provides a common point of integration for all planning, oversight and implementation activities for a project. It provides an interface control mechanism between the software and the hardware on which it runs.

Software configuration management is a support activity which makes technical and managerial activities more effective., and the effectiveness increases with the degree of discipline applied by development and maintenance staff. This true whether software configuration management is administered by a separate software configuration management group, distributed among several projects, or a mixture of the two.

## B.1.2. Software entities managed

Software configuration management includes the control of all of the entities of a product as well as their related documentation. These include the following.

- Management plans.

- Specifications (requirements, design).

- User documentation.

- Test design, case and procedure specifications.

- Test data and test generation procedures.

- Support software.

- Data dictionaries and cross references.

- Source code (on machine readable media).

- Executable code (the run-time system)

- Libraries.

- Databases (data which is processed and data which is part of a program).

- Maintenance documentation (listings, detailed design descriptions etc.)

Software configuration management entities are subject to varying degrees of discipline, depending on the phase. During development, the documentation (requirements specification etc.) are the most import entities to be controlled. During coding, the design documentation needs careful control. When the product is ready for release, it is the source code which needs to be most carefully controlled.

**Note**: Firmware (programs/data which cannot be dynamically modified during processing) is controlled under hardware configuration management.

## B.1.3. Which entities to manage?

The question of which entities are to be managed often arises in the context of what gets captured in each library and when. Consideration should also be given to the hierarchy of entities managed during the process.

The following three-level hierarchy is one way to proceed.

- Configuration item (CSCI, CPCI, system, system, segment, program package, module).

- Component (CPC, CSC, subsystem, unit, package, program function).

- Unit (procedure, function routine, module).

The configuration control boards that are oriented towards business decisions normally selects one level in this hierarchy as the level on which they will control changes. Other levels of the hierarchy would be selected by boards with a different (i.e. technical) focus.

## B.2. Software configuration management process

Since software configuration management takes place within a business environment, to be successful it must blend in with the environment.

Software configuration management defines and implements policies, techniques, standards and tools which help Telstra and its customers by providing effective identification of software, change controls, status accounting, audits and reviews.

The most important activity of software configuration management is managing the change process and tracking changes to make certain that the product configuration is accurately known at all times.

Change management is accomplished by comprehensively identifying each baseline and monitoring all subsequent changes to the baseline. The process applies whether the baseline represents the requirements capture documentation or a fully documented program, including source and object code.

Managing baseline changes requires a scheme for identifying the structure of the software product. The structure is defined by the hierarchical organisation of the product and is extended to include all entities or work-products associated with the program.

The identification scheme must be maintained throughout the life of the product.

As new baselines are created by promotion or release, the aggregate of entities is reviewed or audited to verify consistency with the old baseline. The identification labelling is modified to reflect the new baseline. A record of all baseline changes is maintained.

The diagram below outlines the software configuration management process.

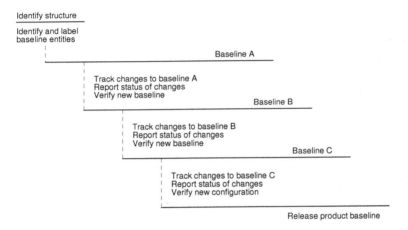

Figure 1: Model of Change Management

## B.2.1. Source code

An important entity to be managed is the source code, since it is the basic representation in readable form of the product being controlled. Other documentation and data are verified against this entity.

Design documentation is verified against the product represented by the source code. The test entities, test data and test reports are used to verify that the executable code matches the documentation. Maintenance and user documentation is also verified against the source code.

Since source code can be interpreted differently by different compilers it is necessary to control the versions of the support software used for a specific released product.

## B.2.2. Levels of control

Levels of control implies all the various forms of control exercised by both management and non management.

Higher management delegates authority to act downwards to include work being done by non management staff. Management also delegates authority to control to non management staff. Authority is the exercising of control over resource allocation. Non management provides the technical data to support in the evaluation. Since the software configuration management plan needs to identify all software configuration items covered by the plan, it must also define the level of management needed to authorise changes to each entity.

In situations where there are many released software configuration items and items under development, there is a need for separate levels of control and authority to approve changes.

## B.2.3. Software libraries

Software libraries enable control and status reporting. Their function is to identify baselined entities and to monitor the status of changes to those entities.

Libraries have traditionally been comprised of a combination of hard copy material and on-line documentation. The trend today is towards all material being maintained on-line in machine readable media, which makes possible the use of efficiency-improving automated techniques.

Being on-line means that the documentation becomes part of the software engineering environment. The software configuration management functions associated with the libraries have to become part of the software engineering environment to make the configuration management process more transparent to the software developers and maintainers.

The number and kind of libraries varies from project to project, according to the access rights and needs of the users (which are directly related to levels of control).

Three types of library are recognised.

- **Dynamic library** (programmers library) - holds newly created/amended software entities. It is used by the programmers and the entities therein are freely accessible to the programmer responsible for the entity.

- **Controlled library** (master library) - used to manage the current baseline(s). It is here that configuration items that have been promoted for integration are maintained. Entry is controlled. Copies can be freely made for use by programmers and others. Changes to the library must be authorised.

- **Static library** (software repository) - where baselines are archived when they have been released for general use.

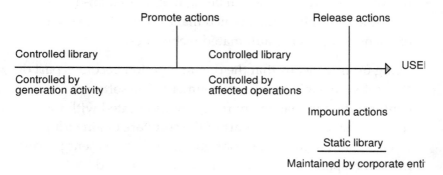

Figure 2: Three types of library

# B.2.4. Configuration control board

Configuration control boards control major issues like schedule, function and configuration of the system as a whole. They work by optimising the change control authorisation process, either by establishing a hierarchy of boards at appropriate levels (i.e program, design etc.) or by a single, overarching board which has control over all levels of the change process.

Configuration control boards are usually made up of senior managers and representatives from software, hardware, test, engineering and support organisations.

A subset is the software configuration control board which handles issues relating to performance, cost, scheduling, common data structures, design changes and the like. The software configuration control board makes the decisions which do not need the attention of senior management.

# B.3. Software configuration management tools

Software configuration management tools must be compatible with the software engineering environment in which they are used. The tools can be categorised as follows.

## B.3.1. Basic tool set

The basic tool set includes the following.

- Basic database management systems.

- Report generators.

- Means for maintaining separate dynamic and controlled libraries. File system for managing check-in and check-out of units, for controlling compilations and capturing the resulting products.

## B.3.2. Advanced tool set

The advanced tool set includes the following.

- Items in the basic tool set.

- Source code control programs that maintain version and revision history.

- Compare programs for identifying (and helping to verify) changes.

- Tools for building or generating executable code.

- A documentation system to enter and maintain specifications and user documentation files.

- A system/software change request/authorisation tracking system that makes requests for changes machine readable.

The advanced tool set is applicable to environments of greater complexity which has more computing resources available.

## B.3.3. On-line tool set

The on-line tool set includes the following.

- Generic elements of the advanced tool set integrated so they work from a common data base.

- A tracking and control system that brings generation, review and approval of changes on-line.

- Report generators working on the common database.

## B.3.4. Integrated tool set

The integrated tool set includes the following.

- On-line tools covering all functions.

- An integrated engineering database with built-in software configuration management commands.

- The integration of the software configuration management commands with on-line management commands for building and promoting units and components.

This set integrates the software configuration management functions with the software engineering environment so that the software configuration management functions are transparent to the engineer. The software engineer only becomes aware of the software configuration management functions when they try to perform an unauthorised action.

## B.4. Software configuration management plan

The application of software configuration management is very sensitive to the context of the project. Thorough planning is necessary for the successful implementation of a software configuration management system. Many of the routine activities associated with software configuration management are straightforward and can be automated quite easily. Effective planning involves working out how the activities are to be performed in accordance with the software configuration management.

The more important planning activities (i.e. a scheme for defining configuration items, change control etc.) are management activities which call for specialised software engineering knowledge and experience.

Software configuration management defines the relationship between activities which span the project development lifecycle. The Software configuration management plan is a central document which brings together the different activities. The cover sheet of the plan is signed by all the people with responsibilities defined in the plan. This makes the plan a living document which is maintained throughout the life of the software.

Maintenance of the plan throughout the life of the software is important as the disciplines of identification, status reporting and record keeping apply throughout the maintenance part of the project.

# B.4.1. Table of contents

The table below outlines the Software configuration management plan. The arrangement of section 3 can vary according to the nature of the software being controlled.

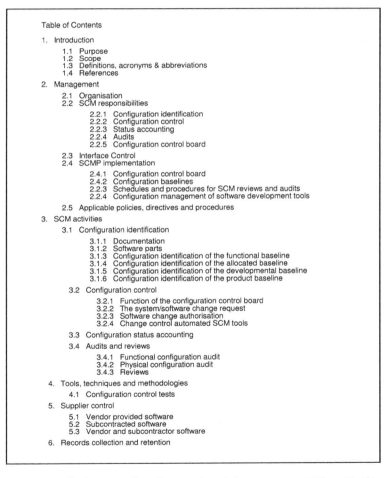

Table of Contents

1. Introduction
   1.1 Purpose
   1.2 Scope
   1.3 Definitions, acronyms & abbreviations
   1.4 References
2. Management
   2.1 Organisation
   2.2 SCM responsibilities
       2.2.1 Configuration identification
       2.2.2 Configuration control
       2.2.3 Status accounting
       2.2.4 Audits
       2.2.5 Configuration control board
   2.3 Interface Control
   2.4 SCMP implementation
       2.4.1 Configuration control board
       2.4.2 Configuration baselines
       2.2.3 Schedules and procedures for SCM reviews and audits
       2.2.4 Configuration management of software development tools
   2.5 Applicable policies, directives and procedures
3. SCM activities
   3.1 Configuration identification
       3.1.1 Documentation
       3.1.2 Software parts
       3.1.3 Configuration identification of the functional baseline
       3.1.4 Configuration identification of the allocated baseline
       3.1.5 Configuration identification of the developmental baseline
       3.1.6 Configuration identification of the product baseline
   3.2 Configuration control
       3.2.1 Function of the configuration control board
       3.2.2 The system/software change request
       3.2.3 Software change authorisation
       3.2.4 Change control automated SCM tools
   3.3 Configuration status accounting
   3.4 Audits and reviews
       3.4.1 Functional configuration audit
       3.4.2 Physical configuration audit
       3.4.3 Reviews
4. Tools, techniques and methodologies
   4.1 Configuration control tests
5. Supplier control
   5.1 Vendor provided software
   5.2 Subcontracted software
   5.3 Vendor and subcontractor software
6. Records collection and retention

Figure 3: Software Configuration Management Plan ToC.

21

# B.4.2. Introduction (section 1)

The purpose of this section of the software configuration management plan is to provide an overview of the purpose and scope of the system.

## B.4.2.1. Purpose (section 1.1)

A brief statement of purpose identifying the system to which the particular software configuration management plan applies. Dependencies on any other software configuration management plans should be noted.

## B.4.2.2. Scope (section 1.2)

The scope identifies the following.

- Specific software configuration management concerns.

- What the plan will and will not address.

- The items to be managed.

- Lowest entity in the hierarchy (control element) that will be reviewed by the top level project or system management configuration control board.

- Smallest useful entity that will be reviewed by technical management.

- Deliverable entities or configuration items to be released as separate entities.

The boundaries of software configuration management activities may also be defined in the scope by means of a diagram.

### B.4.2.3. Definitions, acronyms and abbreviations (section 1.3)

Include all definitions etc. needed to understand the plan or helpful in understanding the plan. This information can be provided by referring to appendix(es) or to other documents.

### B.4.2.4. References (section 1.4)

List the documents referenced in some way by the plan. The documents must already exist - that is, reference shouldn't be made to documents planned to be written in the future.

This information can be provided by referring to appendix(es) or to other documents.

## B.4.3. Management (section 2)

The management section relates the elements of the software configuration management to specific management activities. Specify in this section the budgetary, schedule and resource requirements.

### B.4.3.1.  Organisation (section 2.1)

Allocate configuration management functions to organisational entities. (Interface control matters are handled in section 2.3). Issues to consider include the following.

- What kind of product interfaces have to be supported within the project itself?

- What are the capabilities of the staff who will perform specific software configuration management activities?

- What level of management support is needed to implement the software configuration management?

- Who has the authority to release software, data and associated documents?

### B.4.3.2.  Software configuration management responsibilities (section 2.2)

- Central issues to consider include who has responsibility for the following.

- Configuration identification.

- Configuration control.

- Status accounting.

- Audits.

- Configuration control boards.

Other issues include the following.

24

- Who will be responsible for maintaining the support software. What organisational responsibilities are likely to change during the course of the project?

- What level of management support is needed to implement the software configuration management?

- Who has responsibility for various software configuration management activities?

## B.4.3.3. Interface control (section 2.3)

A common understanding as to the interfaces between the organisational entities involved in a project is a basic prerequisite for a successful project. After reaching an understanding of the nature and scope of the interfaces, document that understanding in this section.

Ask yourself the following.

- What are the various organisational interfaces?

- What are the interfaces between the elements of the computer program?

- What are the dependant hardware interfaces?

- Where is the interface control document?

- What are the procedures for making amendments to the interfaces.

Some interfaces include the following.

- Vendor to buyer.

- Subcontractor to contractor.

- Co-developer to developer.

In addition to the interface details, indicate the following matters.

- How software configuration management disciplines are coordinated and used to manage interfaces throughout the projects life.

- The composition of the configuration control board(s).

## B.4.3.4. Software configuration management plan implementation (section 2.4)

The level of detail for this section depends on the complexity of the system being controlled. For large systems, the following details should be included.

- Configuration control board.

- Configuration baseline.

- Schedules and procedures for software configuration management audits and reviews.

- Configuration management of software development tools.

Indicate how the key software configuration management milestones identified in the plan are to be implemented. Consider the following.

- The prerequisites affecting the plan and the sequence of events in the plan.

- Schedule for accomplishing the events in terms of their relationship with other events.

- Resource requirements (i.e. machine time, disk space specialised tool availability and staff support).

- Are the resources allocated to the project adequate?

- How will the software configuration management activities be coordinated with other project activities?

- How will the different phases be managed throughout the project lifecycle?

### B.4.3.5. Applicable policies, directives and procedures (section 2.5)

Indicate how existing and planned software configuration management policies are to interpreted and applied to the plan. The actual references to these policies should be included in section 1.4 of the plan.

Consider the following.

- Are any standard identification procedures available?

- Are any procedures available for interacting with the dynamic libraries?

- Are any standard procedures available for managing the change process?

- Are any status accounting procedures available?

- Are any audit procedures available?

## B.4.4. Software configuration management activities (section 3)

The software configuration management organisational descriptions outlined in section 2 of the plan describe who has what responsibilities. This section describes how these groups accomplish their responsibilities.

### *B.4.4.1. Configuration identification (section 3.1)*

This section documents an identification scheme that reflects the structure of the product. This is an important task because the flow of management control needs to follow the structure of the software being managed. The main difficulty in defining an identification scheme early in the project is that the detailed structure of the software is rarely known at this stage.

When developing this section, think in terms of what scheme will be used to relate the identification of files to the document-based identification scheme., how is the software identification scheme to be related to the hardware scheme when the programs are deeply embedded in the system (i.e. firmware) and how do we identify programs embedded in ROM?

For large projects, consider the following central issues.

- Documentation - what specifications and management plans need to be identified and maintained under configuration management?

- Software parts - including how do we identify programs which are embedded in ROM?

- Functional baseline.

- Allocated baseline.

- Developmental baseline.

- Product baseline.

## Identify product baselines

Establishing baselines helps to synchronise the efforts of the project team members.

Consider the following matters.

- Do we need more than functional, allocated and product baselines?

- Who authorises any new baselines?

- Who approves a baseline for promotion?

- How and where are the baselines established and who is responsible for them?

- How will the numbering system account for different baselines?

- How are baselines to be verified?

- Are baselines tied to milestones?

Baselines tie documentation, labelling and the program together. They define a product capability associated with performance, cost and other user interests and relate the product to contractual commitments.

## B.4.4.2.  Configuration control (section 3.2)

The configuration control section describes how the software configuration management process is to be managed. The section should identify the procedures for processing changes to known baselines. An appropriate level of authority for controlling the changes needs to be identified for each baseline.

### Configuration control board (section 3.2.1)

This section should  identify the authorities needed for granting change approvals. Section 2.2 of the plan identifies the general roles of each configuration control board. This section needs to go into detail about these roles.

The configuration control board is concerned with managing changes to established baselines of documented configuration items and their components.

Consider the following issues.

- Can the limits of authority be defined?

- Will the project mix programs that are under the control of other configuration control boards?

- Does the project agreement impose requirements on the configuration control board that needs to be reflected in the software configuration management plan?

- How are the different levels of authority determined?

- How are the different organisational bodies phased in when transitioning from one phase of the project to the next?

- How are changes to a baselined product to be batched together for release?

Decision making that affects the allocation and scheduling of development or maintenance resources should be separated from decisions made for technical or marketing reasons.

**System/software change requests (section 3.2.2)**

This section describes the methods used for processing change requests. It should concentrate on the following.

- Define the information needed to approve a change.

- Identify the routing of this information.

- Describe the control of libraries

- Describe or refer to the procedure for implementing each change in the code, in the documentation and in the released program.

The change initiator should analyse the proposed change to assess its impact on all configuration items. The configuration control board should satisfy itself that a thorough assessment has been done.

The documentation hierarchy should be fully defined and a change to any level should be considered to determine any impact on higher levels of documentation and that the change has filtered down through the lower levels to be implemented in the code.

# Software Configuration Management

When processing changes, consider the following issues.

- What is the information necessary for processing a software/system change request?

- What kind of information will a configuration control board need to make a decision?

- What software configuration management support is provided by automated tools available in the environment?

- Will changes in procedures be required to support different kinds of reviews during each of the phases of the lifecycle?

- Is there a need for dynamic libraries and controlled library interfaces?

- Is there a need for controlling all access to a library or just controlling changes made to the information in the libraries?

- Does the library system provide for an audit trail of change histories?

- Are backup and disaster files taken into account?

- Are there archive procedures to provide support to the static library for the life of the project?

- How are source code items associated with their derived object (executable code) programs?

- How does the change process itself support or accommodate the development of new versions or revision?

**Software change authorisation (section 3.2.3)**

The levels of authority needed to make changes to configuration items under software configuration management control can vary. The system itself can dictate the level of authority needed. For example, internally controlled software tools will probably need fewer controls than man-rated critical software.

The basic issues to consider when defining levels authority are as follows.

- Is the level of authority consistent with the identities identified in section 3.1 of the plan.

- What are the levels of control assigned to the modifiable units of the software during top level and detail design stages for developmental baselines?

- Do control levels for developmental baselines need to be reviewed by management?

- Are there significant increases in levels of control for transitions between developmental baselines?

- Does management need to know who exactly requested a change?

- Do changes originating from outside the organisation require different authority for approval; than changes from within the development group?

- Is there sufficient flexibility to adjust the levels as the project proceeds.

The plan needs to clearly define the level of authority to be applied to the baselined entities throughout the life of the project.

## Change control automated software configuration management tools (section 3.2.4)

This section describes the support software used in the building and maintaining of the product throughout its lifecycle. The focus is on describing the controls used to manage the support software.

Support software can be user-supplied, developed in-house, leased from a vendor or purchased off-the-shelf. The developer needs to ensure that the support software is available for as long as necessary (for example, compilers must be archived for later use *as is* when making improvements later to prevent subtle compiler dependencies from turning a simple enhancement into a major upgrade).

Consider the following issues when planning the software configuration management of support software.

- What is the total set of support software used to design, develop and test the software controlled under the software configuration management plan?

- Is this set of software archived and maintained for the full lifecycle of the computer program products?

- What procedures are to be followed when introducing new versions of support software within the scope of the plan?

- How is the hardware configuration used to develop and maintain the software product identified and maintained for the full life of the product?

It's important to determine the appropriate level of software support needed for the maintenance of the product throughout its full lifecycle. That is, what is sufficient for the job but not too costly.

When a product baseline is established it is important to archive all environment and support tools along with the production code.

## B.4.4.3. Configuration status accounting (section 3.3)

This section identifies what information is needed for various activities, how to obtain that information and then report on the information. The emphasis is on getting the right information at the right time so as to be able to produce the right report when it is needed.

Status accounting equates to financial accounting in that it tracks the flow of software through it's development in much the same way as the flow of funds are tracked through a chart of accounts.

Following the analogy, several accounts can be established for each configuration item. Individual transactions are then tracked through these accounts as and when they happen.

The status accounting system, at its minimum reports the transactions occurring between software configuration management controlled entities.

The functional capabilities of the library system and nature of the software configuration management tools largely determines the capabilities of the status accounting function. As well as providing live information about the development process, the configuration of each released baseline needs to be documented, together with the exact configuration of the released system.

When planning for status accounting, consider the following matters.

- What type of information needs to be reported?

- What degree of control is needed by the customer?

- Who are the audiences of the reports?

- What are the formalities for requesting reports?

- What information is needed to produce reports?

- In the case of large systems, is there a need for handling rollover of identification sequences?

Status accounting data is useful for the project manager to keep track of progress.

## B.4.4.4. Audits and reviews (section 3.4)

The audits and reviews section of the software configuration management plan describes the procedures used to verify the software product (executable code) matches the configuration item descriptions in the specifications and documents, and that the package being reviewed is complete.

As a general rule, the organisation doing the quality assurance functions should also do the audits which address change processing functions, the operation of the software libraries and other activities related to software configuration management. This is in contrast with the reviews and audits done within the scope of a software configuration management activity that verify that a software product is a consistent, well-defined collection of parts.

The audit is a way to check that the developers have done their job completely and in a way that will satisfy external obligations. Anomalies detected during the audit must be fixed, as must the root cause of the problem be fixed to avoid the problem from recurring.

**Functional configuration audit (section 3.4.1)**

The functional configuration audit involves determining that all items identified as being part of the configuration are actually present in the product baseline to check that it conforms with the functional requirements specification. Both functional and physical audits provide notice that contractual obligations are nearly complete and to give the customer the evidence that the product is ready to be put into operation.

Issues relating to both kinds of audit are as follows.

- Do we need to have more than one audit per baseline?

- Is there a single, separate audit trail for each component and for the personnel working on them?

- Are provisions made for auditing the software configuration management.

**Physical configuration audit (section 3.4.2)**

The physical configuration audit, like the functional audit also checks that the nominated items are present. The audit must also establish that the correct version and revision of each part are included in the product baseline and that they correspond to information contained in the baseline's configuration status report. See the above list for issues relating to this kind of audit.

**Reviews (section 3.4.3)**

Management and technical reviews are held at regular intervals throughout the project.

# B.4.5. Tools, techniques and methodologies (section 4)

The tools, techniques and methodologies section outlines how to make the plan work. A well-planned project makes use of good planning tools.

The audit trail reports should point towards the milestones on the planning charts which gives management an effective way of tracking progress. An automated software configuration management system should include some way of integrating these planning tools with the software configuration management database to provide all concerned with an on-line tool for creating products and checking their current development status in real-time. The correlation of the tools against the plan produces quantitative performance and schedule figures,

The tools, techniques and methodologies used to implement a software configuration management are described in terms of a set of libraries and describes how data is captured and stored in these libraries. When planning the tools, techniques and methodologies section of the software configuration management plan, consider the following.

- What are the number and type of libraries to be established (i.e. a dynamic, controlled or static libraries)?

- How much change activity is anticipated on the project?

- Can the software configuration management tools in the library be used to manage documentation and source code?

- How much training will be needed to implement the tools?

## B.4.5.1. Configuration control tools (section 4.1)

Particular tools as follows.

- Source management system - a file system for checking vendor-supplied products and internal software.

- Package management system - a vendor supplied tool which generates software.

- Systems/software change request tool.

- Software change authorisation

- Status accounting report generator

Supplier control (section 5)

Section five of the software configuration management focuses on how to place effective configuration management on the software elements over which you have no direct control.

Suppliers fall into two categories - subcontracted and vendor.

Issues to consider include the following.

- Is the product to be used internally, delivered as part of the finished product or both?

- What post-delivery error-correction arrangements are required?

- What changes are the supplier entitled to make after delivery?

- When should audits be performed?

- Is there a need to pass through software configuration management tools to a supplier or a vendor?

- What periodic reviews of the subcontractors work are necessary?

### B.4.5.2.  Vendor provided software (section 5.1)

Warranties contained in purchase orders may be difficult to enforce. The specific criterion is that the vendor should furnish the computer program media as specified by a purchase order or by the suppliers documentation referenced in the purchase order.

Issues to consider are as follows.

- How is the vendor software supplied identified?

- How are license agreements and data rights protected and enforced?

- Are there limitations on duplicating the documentation or on the customer making copies of the program?

- How will vendor support be provided over the life of the product?

- How will copyright interests be protected?

- How will legal copies of leased software be controlled?

## B.4.5.3. Subcontracted software (section 5.1)

If some part of a software development project is subcontracted to another organisation, the responsibility for software configuration management is passed to that organisation. This notwithstanding, the vendor is only responsible for their portion of the total product, not for the integration of that portion with the whole.

- The integration of a subcontractors software configuration management could be arranged as follows.

- Provide a library management tool and monitor its use.

- Allow the subcontractor to promote code to your software generation system and controlling it in the same way as your own code.

- Obtain the source of all subcontractor deliveries and recompile/relink it using the buyers software generation tool.

Consider the following factors when defining subcontractor relationships.

- What software configuration management concerns need to be added or removed from the contract?

- Who is responsible fro auditing versus enforcing software configuration management for contractual products?

- What audits and procedures need to be established where the subcontractor has no documented software configuration management practices or procedures?

### B.4.5.4. Vendor and subcontractor software (section 5.3)

Evidence must be sought that all vendors and subcontractors are audited for compliance with good business practice software configuration management. The frequency and method of the audits are determined by the size, criticality and dollar value of the project.

## B.4.6.  Records collection and retention (section 6)

The concept underlying this area is that information is kept only for as long as it is needed, and kept in a way that details may be retrieved without delay. Good configuration management involves keeping copies of release material for backup and disaster recovery. In some case, it is worth retaining test records where it may be necessary to prove that proper testing has been done.

Consider the following issues.

- What kind of information needs to be retained?

- What data needs to be retained for trend analysis?

- Is everything necessary to recreate the system available on archive?

- Is the media protected from disaster?

- Is it necessary to keep copies of software which has been licensed for use and distribution?

- What activities need to be recorded for maintaining a product after the project is completed?

- For whose use are the records being maintained?

- How are the records to be kept (i.e. media, location)?

- How long will data be kept?